Developmental Universality and Unity of the Universe

By

Vladimir Groh

To the image of God He created...

Table of Contents

Introduction

Einstein created Relativity Theory which describes the Universe constitution through mathematics; it does not, however, address the physic nature of things. The theory of General Relativity, in contrast to the theory of Special Relativity, does not correlate in principle with another recognized theory — the Quantum Mechanics which outlines microcosm processes. Criticism evolved as soon as General Relativity had appeared; however, no well-founded independent alternative to it was presented.

The only fact that has been universally recognized so far is that an alternative to General Relativity shall qualify the most strict requirements — it shall be reduced to Special Relativity at the rates that are proximal to velocity of light and a low gravity; it shall outline General Relativity effects equally well as Special Relativity does at high gravity; it shall be reduced to Newton's mechanics at low gravity and small velocities.

Creating a universal theory that would combine both Newton's and Einstein's theoretical calculations are perceived as a task difficult to accomplish. And the ambiguity of their scientific approaches to gravity is only one of many things that can prove it. In accordance with Newton, gravity had been described as a field-force interaction, i. e. each object that has a mass creates the force-gravity field by its attractive power; this field influences other objects with a mass. It had been also assumed that gravity was a force. There seemed to be no doubt about such argument until the proof was given that light being only an electromagnetic wave, i. e. a field structure without a mass, turned aside from massive bodies. Therefore, Newton's statement that "no mass means no force" had come under criticism.

After Einstein studied in detail the particularities of gravity development, he first disputed that the gravity should be categorized as a force; then he introduced the concept of space-time curvature near gravitational masses.

This example allows making a reasonable conclusion — it does not seem possible to combine Newton's and Einstein's theoretical computations in a universal single theory. However, in our beliefs, the truth shall be somewhere in between. And this "golden mean" which locates at the interface of these two scientific approaches can be the effects that the Einstein's theory of General Relativity had foreseen and the reasoning of them was given from the Newton's point of view who was an strong supporter of a model of the ether.

We believe that if Newton had studied the ether environment in detail taking into account all its properties, then, very likely, most of his scientific statements including the theory of gravitation would take a bit different shape.

1. Levels and Properties of Ether

This is why we suggest analyzing the ether itself in the first place and its properties upon interactions. When looking at many phenomena and processes that today do not seem to be well-founded or cogent, this order of study helps to see them from a new perspective. Let us get clear with the ideas:

Firstly, there is the material level which is a combination of elementary particles — electrons, protons, neutrons, chemical elements, molecules, and macro bodies.

Secondly, the photon level of the ether consists of photons that have, by analogy with atoms has a complex structure and interact in creating photon field. The equally charged photons push off from each other.

By assuming the existence of the photon field we can explain different interactions including relativistic effects. Let us adduce the following proofs of the existence of photon field:

1. Isotropy of radio-frequency radiation from a point source.

2. The deviation of a light ray from a star near the Sun can be explained by the density of the photon field at different distances from the Sun, i. e. a regular ray refraction takes place in environments with different density.

3. Mass defect upon nuclear reactions is a part of the mass that has fitted into another level (the photon field) and changed to the electromagnetic radiation. Mass defect can be to the both sides, so there is no need in Higg's bosons to explain how the mass emerges from the photon field:

4. $\Delta E = \Delta m\ c^2 \rightarrow \Delta m = \Delta E/c^2$. The mass change sign (+ or -) depends on whether the energy is emitted or absorbed.

5. Displacement currents evolving in electromagnetic radiation cannot be explained without the existence of the photon field.

6. The velocity of electric current distribution is equal to the velocity of electromagnetic wave propagation within the photon field rather than the velocity of motion electrons in metal that equals to 0.5 m/sec.

7. Continuous growth of the masses of all space objects can be explained by the continuous matter fusion from the photon field.

8. Availability of the photon field ensures that empty space curvature can be set aside and the space bodies interaction can be described by the same formulas of dynamics but in the curved photon field.

Thirdly, the gamma photon level of the ether consists of gamma photons that are much smaller than the photons and have a complex structure and create the gamma photon field. The equally charged gamma photons push off from each other.

The frequencies and seizes of the lengths of the waves of the photon and gamma photon levels are shown in Fig. 1.

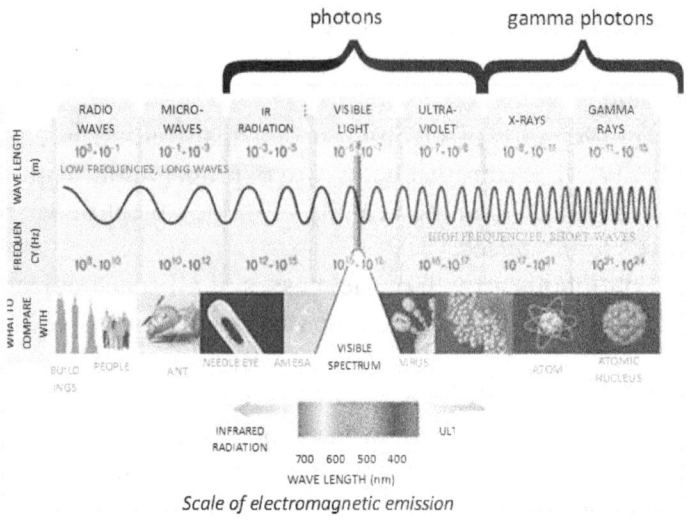

Scale of electromagnetic emission

Figure 1. The frequencies of the levels

It should be realized that the boundaries of the levels of the ether are relative and that the other levels with smaller objects can exist outside the gamma photon level.

It is known, that the highest energy, in accordance with the Planck's law, can be transported by the particles with a higher vibration frequency. In addition, they are more energetically dense because they are, as we assume, "building blocks" for the photon level. It is the gamma photons that determine the main energy of the Sun quantity we have extracted because the particle energy can be defined by the frequency of its vibration: $E = h\,v$, and the frequency of the gamma photons is 10^{10} times higher than the frequency of the oscillation of the photons.

Of all three levels that we have taken for reviewing herein (there are much more of them in reality), the gamma photon level is seen as more energetically dense one because it is the finest level that can be "packed" with smaller voids, and therefore has own variations of a higher frequency and, as previously mentioned, is the "building block" for the next level — the photon level.

Conclusions:

1) the finer levels of the ether are the matrices that create more rough levels from our perspective;

2) all levels have the same charge sign which prevents the agglutinating of their particles;

3) we position the finer levels as the fields towards to the particles of the material level that are the energy sources.

2. Synthesis Is the Principle Factor of the Development of the Universe

Before we move on to the issues of the space interactions and gravity, we should first study fusion, which is one of the principle ideas from our perspective.

It is assumed that the fusion is a process where a more complex matter develops from a simple matter and that the mass defect of the source materials and the energy emission follow such development.

The most obvious example of this effect is the thermonuclear reaction of helium / hydrogen fusion.

Upon the fusion that is followed by the mass defect and the local reduction of quantities in the ether medium, a local reduction of the surrounding field potential regularly occurs. The difference in potentials that occurs in the photon field near the fusion center leads to the medium flow, and, subsequently, to a deformation of the field.

By the mass defect is meant a difference of the actual and virtual masses followed by the change of the part of virtual mass to the energy of frequency range appropriate for a finer level. Virtual mass of the photons moving to the fusion center is a growing function dependent only on the distance to the fusion center. During the fusion, the virtual mass transforms into the material mass.

It should be noted that as soon as the process of the local fusion related to the mass defect and the reduction of the source capacity of components ends, a step-like reduction of the potential pressure shall occur in this local capacity. This pressure reduction in the specified quantity will be followed by the slowed down fusion until next increase in the value of such pressure to the "threshold" value specific to the initial fusion behavior. The fusion process is constant and becomes discrete at the border of the two levels — the photon level and the material level — with the transmission of one into another followed by the defect of masses.

Photon fields create fluxes that are heading to the fusion area and are moving in these directions with a variable acceleration.

In accordance with Bernoulli's law, the fact that the pressure of the flowing media is reduced back by quadratic dependence at an increased flow rate is correct to describe the dependencies at material media flow and this is also true of the photon field. It should be noted the fact that as the photon flux approaches nearer the fusion center, the total area of the flux is reduced by quadratic dependence, and this shall logically lead to a significant increase in pressure by quadratic dependence. Therefore, the slight decrease in pressure of the flowing medium caused by the increase in the flow rate by quadratic dependence is excessively compensated by the increase in pressure by quadratic dependence which is due to the change in the geometrical size of the flux. In view of the foregoing, the density of the photon flux significantly increases as it approaches the fusion area.

So we obtain as follows:

1. The accelerating photon field with a vector to the center of the fusion (e. g. the Earth) creates gravity in interaction with material objects.

2. The continuously growing mass of space (gravitating) objects, depending on the degree of their deformation, e. g. the Earth is the deformation of the photon field, and the Sun (star) is the deformation that affects also the gamma photon level.

9

The conclusion: The fusion in the ether medium causes its deformation.

We will cover in detail the issues related to gravity in the next chapter, and now we will consider the conditions that contribute to the formation of a focal point at the fusion initial stage (singularity). If we study the propagation of electromagnetic disturbance in the ether medium rather than in the vacuum then the magnetic field, which works as a damper limiting the rate of light propagation speed, can create the conditions for the local energy accumulations. These areas combined with the potential internal pressure can be exposed to electromagnetic disturbances from several directions both from different sources and from a single source by altering the disturbances directions with the possible manifestation of resonance phenomenon. In the real life, the quantities of the photon field can receive the disturbance pulses at different angles, of different intensity, and so on, and, obviously, some of the quantities can twist under the influence of the disturbance vectors to accumulate even more energy. Creating the conditions for the point-like fusion initiation. Let us study how the "non-material" photons that have no mass can generate material particles. For review, let us give an obvious example that shows how two effects, unrelated at first sight, are actually a reasonable logical relation of a single process. As the example, we refer to the calculation of the mass change in the photon as it moves from the Earth geostationary orbit to the center of the Earth.

The mass of green (medial) photon: $m_g = 8.8 \cdot 10^{-36}$ [kg]. (Fotony i struktura fotonnogo polya [Photons and the Structure of the Photon Field] [In Russian], p. 4. Chelyabinsk, Dom pechati, 2010.)

Geostationary orbit radius: $R_g = 3,600,000$ [m].

10

With m — mass moving to the Earth from h — the height, a transition of potential energy into kinetic energy occurs:

$$W_p = m_g \cdot g \cdot h \quad (1)$$

$$W_k = m_g v^2 / 2 \quad (2)$$

This energy W_p is transformed to the photon mass increase Δm. Using the Newton's law of universal gravitation, let us set down the equation to calculate the gravitational force that affects the material body located on the geostationary orbit:

$$F = \gamma \cdot m_1 \cdot m_2 / R^2 \quad (3)$$

where: $m_1 = 5.9742 \cdot 10^{24}$ kg – the Earth mass;
$\gamma = 6.67 \cdot 10^{-11}$ $m^3 kg \cdot f^{-2}$ – gravitational constant

For the equation (3):
$$F_1 = 6.67 \cdot 10^{-11} \cdot 5.9742 \cdot 10^{24} \cdot 8.8 \cdot 10^{-36} / (36 \cdot 10^5)^2 =$$
$$(352.176/1,296) \cdot 10^{-33} = 2.7 \cdot 10^{-34} N.$$

The force that influences the photon at the Earth center can also be determined using the equation (3):

$$F_2 = 6.67 \cdot 10^{-11} \cdot 5.9742 \cdot 10^{24} \cdot 8.8 \cdot 10^{-36} / R^2$$

additionally, R equals to the photon radius: $r_g = 2.745 \cdot 10^{-12}$

$$F_2 = 6.67 \cdot 10^{-11} \cdot 5.9742 \cdot 10^{24} \cdot 8.8 \cdot 10^{-36} / (2.745 \cdot 10^{-12})^2 =$$
$$(352.176/5.49) \cdot 10^{-12} = 32.07 \cdot 10^{-12} N.$$

Let us determine the acceleration that has the photon at the Earth center:
$$a_2 = F_2 / m = 32.07 \cdot 10^{-12} / 8.8 \cdot 10^{-36} = 3.65 \cdot 10^{24} \ m/sec^2$$

Let us determine the increment of the photon mass:

$\Delta m = m_g \cdot a_{medial} \cdot h = 8.8 \cdot 10^{-36} \cdot (3.65/2) \cdot 10^{24} \cdot 36 \cdot 10^5 = 6 \cdot 10^{-5}$ kg.

This value of the photon mass increment, which is many times greater than the electron mass, can be determined as $(9.1 \cdot 10^{-31} kg)$, that proves that the fusion of the electrons from the photons shall initiate in the planet interior rather than exactly at its center.

Additionally, the density of ρ — the pressurized photon will be:

$$\rho = (\Delta m + m)/(4/3)\pi r_g^3,$$
$$\text{where } r_g = 2.74 \cdot 10^{-12} \, m,$$

$\rho = (6 \cdot 10^{-5} + 8.8 \cdot 10^{-36})/(4/3)\pi(2.74 \cdot 10^{-12})^3 = 0.067 \cdot 10^{41} = 6.7 \cdot 10^{39} \, kg/m^3$.

The obtained quantity corresponds to the density of the neutron or the proton. Therefore, the nucleons are synthesized from the photons at the Earth center.

This example not only proves the increment of the object mass upon the increase of its displacement velocity but also it gives a theoretical justification to the fact that the elementary particles would have to be synthesized from the photon field at the Earth center.

We have already noted above that the increase in the planetary sizes is, most likely, related to the fusion processes that take place in their interior, and now the perquisites of their presence become more obvious.

In our belief, any particles with a complex structure and not only the material particles are able to be synthesized at a specific excess pressure, i. e. they can develop into new stable objects with specific parameters, with fewer quantities, and with the emission of some quantity of energy into another level.

Conclusion: We have studied the fusion which evolves and progresses in the medium (in the ether) and have proved that:

1) There are areas where the conditions for the point fusion initialization evolve.

2) Upon the fusion, a mass defect evolves — it is the particles of a finer level with a high energy that create a super hard core because the process flows at the surface of the ball pressurized by the fusion products.

3) There are the conditions when the non-material photons receive the mass and energy comparable with the material particles and create the matter.

4) The medium fusion, where, at least, 3 levels of the ether are affected, deforms it, thus creating the interaction laws of the material world.

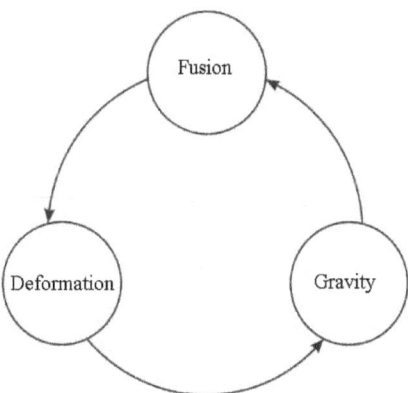

Figure 2. No fusion — no gravity

It is believed, that the Sun is a thermonuclear reactor where the reactions of helium hydrogen fusion, lithium helium fusion, etc. up to the iron fusion are present; then the fusion is

13

accompanied not only by the energy emission but rather by its consumption from the environment. If we calculate on the basis reasoning from this model to determine the time all the hydrogen (equal to the Sun mass) needs to transmit to iron, then the Sun would exist only for 100 million years. Where does the energy, which makes the Sun emit photons, various energies, elementary particles, gases, and larger chemical elements for 4,5 billion years, come from? It seems that the source of such energy is the fusion at a finer level when photons are synthesized into electrons, protons, neutrons, and only then into hydrogen, etc. following the common scheme.

This can be schematically presented in Figure 3.

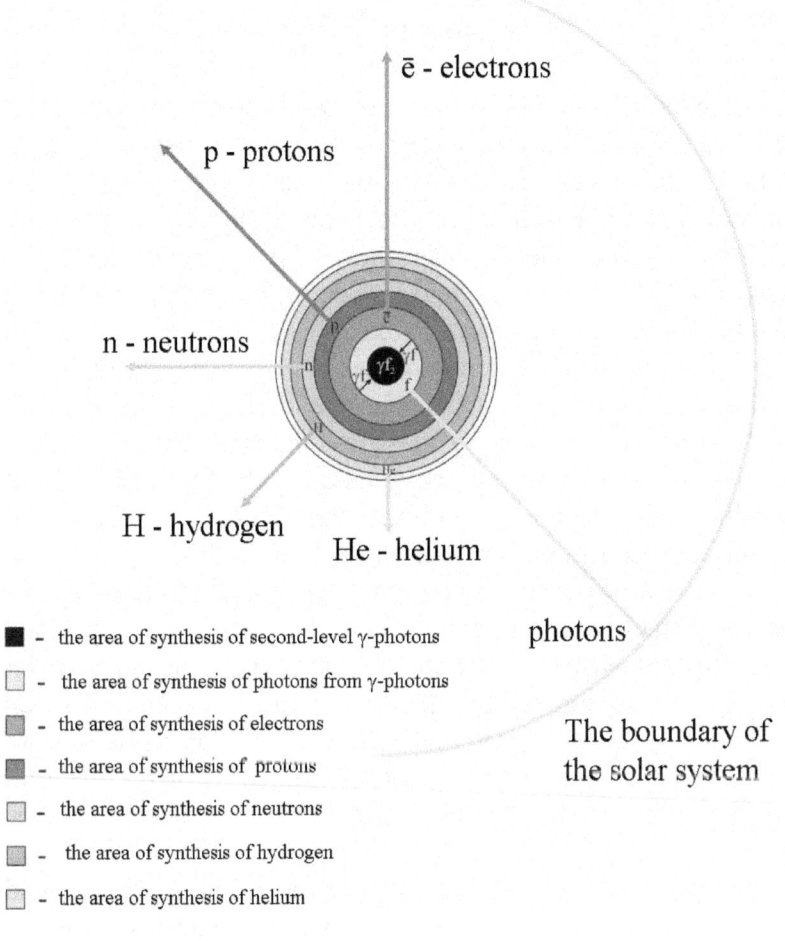

ē - electrons

p - protons

n - neutrons

H - hydrogen

He - helium

photons

■ - the area of synthesis of second-level γ-photons

□ - the area of synthesis of photons from γ-photons

▨ - the area of synthesis of electrons

▨ - the area of synthesis of protons

▨ - the area of synthesis of neutrons

▨ - the area of synthesis of hydrogen

□ - the area of synthesis of helium

The boundary of
the solar system

Figure 3. The fusion areas in the Sun body

3. A Note on Gravity

As we have mentioned above, the near-Earth photon field is accelerating with the vector directed to the Earth center as a result of the fusion. It follows that the elementary particles composing the material objects located in this photon field are interacting with the accelerating photon field and the gravitational force evolves, and its vector matches the vector of the photon field. According to the above mentioned, we can study gravity as an interaction of the objects through the medium that itself takes part in energy and matter conversions.

The concept of the body mass is a reasonable place to start the study of gravity. In our understanding, the body mass is the quantity of the neutrons, protons, and electrons in this body and it manifests itself in mechanical interactions.

The inertial mass of the body can be also determined by the number of the neutrons, protons, and electrons in this body and it manifests itself only in attempting to change the body travel velocity in relation to the photon field; i. e. inertia is an interaction of the body moving with acceleration in relation to the surrounding photon field. The inertial mass of the material body manifests itself regardless of the presence of other bodies nearby.

Gravitational mass is a set of neutrons, protons, and electrons in the specified body quantity and it manifests itself in the interaction with the moving photon field respective to the presence of other material bodies in it.

Everyone knows from their schoolroom days, that the most visual way to demonstrate the gravity forces and free-fall acceleration is the Newton's tube; this device is a 1 meter glass tube which has one end welded up and the other end equipped with a valve. To perform the demonstration experiment, three different objects are placed into the tube, e. g. a shot, a cork, and a feather. Then the tube is turned upside down very quickly. All three bodies will drop to the bottom of the tube but at various times — first, the shot, then the cork, and finally the feather.

This happens if air is present in the tube. But as we evacuate air from the tube with a vacuum pump and turn it upside down, we can watch how all three bodies reach the bottom of the tube at the same time. Air evacuation from the tube has resulted in the elimination of the resistivity of air for all three bodies, regardless of their weight and volume, so that they were affected only by the gravity force with the free-fall acceleration similar for all bodies.

The examination of the above experiment gives us, additionally to the described above, the note that this experiment is also the most visual way to demonstrate the transportation of the photon medium at the Earth surface. Air evacuation from the tube did not affect the transportation of the photon medium within the medium and the interaction of the photon flux with the electrons, protons, and neutrons of the free-falling bodies.

The elementary particles of some material bodies can not interact with molecules at their best. Accordingly, air evacuation from the tube has eliminated the slowing-down effect of the air with the various force on the different objects; this has enabled the creation of the most favorable conditions to secure the purity of the experiment conducted upon determination of the effects of the flux of the photon medium on the elementary particles that combined in various ways in building of the specific material bodies, regardless of what kind of body it was — the shot, the cork, or the feather.

In this case, they all move similarly in the general photon flux regardless of their shape and sizes, just like a log and a match would move similarly in a general river flow.

So, we assume that, among other matters, the experiment that uses the Newton's tube with the evacuated air can qualitatively illustrate not only the movement of the photon medium towards the center of the Earth (planet) but also allows us to make quantitative calculations of its acceleration, which is approximately equal to 9.81 m/sec^2, e. g. for the Earth.

The free-fall acceleration can be approximately calculated (in m^2/sec) by the empirical formula:

$g = 9.780327(1 + 0.0053024 \, Sin^2\varphi - 0.0000058 \, Sin^2\varphi) - 3.086 \cdot 10^{-6} \cdot h$,

where φ – the latitude of the location under consideration, h – the height above seal level in meters

The obtained value matches only approximately the free-fall acceleration at this location.

The quantitative values of the acceleration at the various height h above sea level are shown in Table 1.

Table 1

h, km	0	5	10	15	20	50	100
g, M^2/ sec	9,806	9,791	9,775	9,760	9,745	9,654	9,505

h, km	500	1000	10^4	$5 \cdot 10^4$	$4 \cdot 10^5$
g, M^2/ sec	8,45	7,36	1,5	0,125	0,002

The values in the table give evidence that the free-fall acceleration value is not constant but is increasing as it approaches the center of the Earth. This fact also confirms the relevance of the above calculation and the proposed assumption regarding the reasoning behind the fusion of the electrons from the photons in the Earth interior.

From these, we can easily understand what the gravity force represent. The force is known to occur when the mass of the body and the acceleration are present. In this way, any body, which has a mass and is located near a planet where the photons are synthesized (including the Earth), will be, at the same time, automatically present in the accelerating medium directed towards the center of the planet.

By multiplying the mass of the material body under consideration by the acceleration of the photon medium, which we customary call the free-fall acceleration, we will obtain the force at which the body is held against the planet by the photon flux. This force is also known as the gravity force.

The fullness and integrity of the scientific approach that we propose and that is based on the analysis of the effects, processes, and interactions in the ether medium shall be complete with the reasoning for the attraction of two bodies at the Earth surface.

Using the Newton's law of universal gravitation, let us set down the equation to calculate the gravitational force that affects the material body:

$$F = \gamma \cdot m_{Earth} \cdot m_2 / R^2 \qquad (1)$$

where: $m_{Earth} = 5.9742 \cdot 10^{24}$ kg – the Earth mass;
$\gamma = 6.67 \cdot 10^{-11}$ $m^3 kg \cdot f^2$ – gravitational constant
m_2 – the mass of the body.
R = 6,371,302 m – the mean Earth radius

At the other hand, the two bodies are attracted to each other, probably, according to the same law, see Figure 4.

19

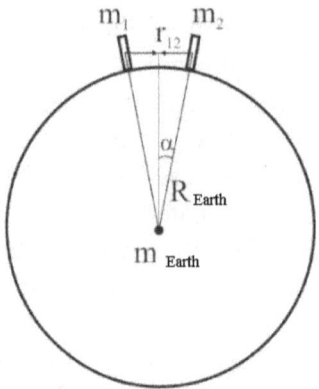

Figure 4. The attraction of two bodies

Let us assume $m_1 = m_2 = 100$ kg, $r = 1$ m.
$$F_{Earth1} = F = \gamma \cdot m_{Earth} \cdot m_1/R^2 = F_{Earth2}$$
$$F_{Earth1} = 6.67 \cdot 10^{-11} \cdot 5.9742 \cdot 10^{24} \cdot 100/(6{,}371{,}302)^2 =$$
$$= 39.847914 \ 10^{15}/40{,}593{,}489{,}175{,}204 = 986 \text{ N}.$$
$$F_{12} = \gamma \cdot m_1 \cdot m_2/r^2$$
$$F_{12N} = 6.67 \cdot 10^{-11} \cdot 100 \cdot 100/1^2 = 6.67 \cdot 10^{-7} \text{ N}.$$

At the other hand, the force F_{12} can be determined as a pressurizing force from the two vectors of the attraction force of these bodies by the Earth.

$$F_{12pr} = F_{Earth1} \cdot \sin\alpha$$

$$\sin\alpha = (r_{12}/2)/R$$

$$\sin\alpha = 0.5/6{,}371{,}302 = 7.84768953 \cdot 10^{-8}$$

$$F_{12pr} = 986 \cdot 7.84768953 \cdot 10^{-8} = 7.74 \cdot 10^{-5} \text{ N}.$$

Accordingly, the pressurizing force F_{12pr} from the two vectors of the attraction force of these bodies by the Earth is a hundred times higher than the force of attraction between the two bodies F_{12N} .

The example soundly shows that the nature of the initiation of the attraction between bodies in the field of the gravitational object can be determined quite logically by the theoretical prerequisites and, as a result, by the "space curvature".

4. Building the Model of a Part of the Universe

The Universe is not static and it shall be regarded only in light of its dynamic modifications that are related to the energy exchange between the various levels and that are followed by the formation and destruction of the objects in the space.

The evolutionary vector of the Universe is directed from the finer levels to the material levels (in our understanding).

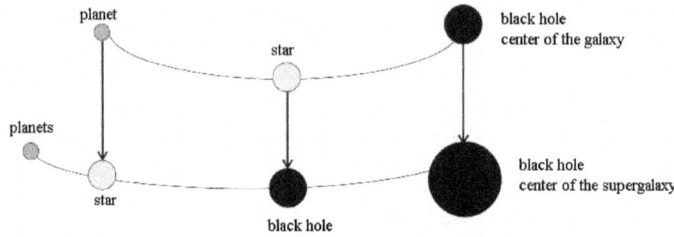

Figure 5. The evolutionary vector

The whole process under consideration related to the co-presence of the photons and the gamma photons in the same space shall be studied as a closed system pressurized on all sides rather as an open system where the densities in the gas-like, liquid, and dry substances are distributed in accordance with the principle that more dense are located at the bottom, and less dense are located at the top. This process is much alike the carbonation of water where the water (in our case it is the gamma photon level) with a lower density as compared to the gas density is in a closed volume and the carbon dioxide with a lower density is added to it under pressure. In addition, the gas molecules (of the photon level in our case) are built in between the water molecules in the limited volume of the vessel (in our case, of the cosmic space). Upon the carbonation of the water, the means of creating pressure of the carbon dioxide which is being built in the water structure is the compressor (the gas cylinder under high pressure, etc.), while, in case when the photon level is being built in the structure of the gamma photon field, the star (including the Sun) which produces the photons from the gamma photons shall be such means.

Using the above assumptions let us proceed directly to the building the Physical Model of a Part of the Universe:

1. Let us assume the limitation of the photon propagation by the sphere around the Sun with the radius equal to the radius of the solar system. Let us denote the set of the photons with the same charge as the photon field.

2. Let us place the stars — the planets — into the photon field. We will consider them as the fusion centers of matter from the photons in the photon field.

3. Accordingly, let us place stellar systems into γ–photon field of the Galaxy with the "black hole" as the center. Let us assume that the black hole is the source of the γ–photon field within the Galaxy.

The fusion processes are similar at the level of the Galaxy, the star, and the planet. Yet there are differences in:

a) the fusion subject: the black hole produces γ–photons from the second level γ–photons (γf_2) which corresponds to γf_2 – the field. The stars produce the photons from the γ–photons. The planets produce the matter from the photons and the photon field of the star;

b) the depth deformation of the ether:

within the Galaxy, the fusion affects the γf_2 level, the γf level, and the f level;

23

in the stellar system, the fusion affects the γf level, the f level, and the material level;

in the planets, respectively, the f level and the material level;

c) the fusion products:

e. g. the Sun produces photons, electrons, protons, neutrons, H_2, He, and etc. up to the Fe fusion.

The planets produce electrons, protons, neutrons, H_2, He, and etc. up to the trans-uranium elements.

It should be noted that the fusion is followed by the energy emission (the mass defect) in a denser level:

for the black hole, it is the γf_3 level;

for the star, it is the γf_2 level;

for the planet, it is the f level.

d) the rate of the processes.

Example: The sound velocity in air (the material level) – 330 m/sec; the velocity of light in vacuum (the photon level) – $3 \cdot 10^8$ m/sec; let us determine the velocity of interaction in the γ–photon level. Based on the equation $\lambda = c' / v_p$, we can calculate the rotational frequency of the photons in the electron:

$v_f = c / 2\pi r_f$

$v_f = 2.9975 \cdot 10^8 / 6.28 \cdot 0.8 \cdot 10^{-17} = 5.966 \cdot 10^{24} \ [c^{-1}]$

When we know the rotational frequency of the photon in the electron, i. e. the frequency of interaction in gamma photon level, we can determine the velocity of the wave propagation within this level:

$$c_\gamma = c \cdot (v_f / v_\gamma) = 2.9975 \cdot 10^8 \cdot (5.966 \cdot 10^{24} / 3 \cdot 10^{15}) = 6 \cdot 10^{17} \ [m/s]$$

(in the γf level, the velocity exceeds the velocity of light, i. e. the velocity of moving of the γ–photon field exceeds the velocity of the wave in the photon field; this explains why the black holes cannot be seen).

It would be reasonable to study the stellar system as the unified object. In addition, it should be borne in mind that the Universe is not static and it shall be regarded only in light of its dynamic modifications that are related to the energy exchange between the various levels and that are followed by the formation and destruction of the objects in the space.

Let us place the stellar systems, including our Sun, into the γ–photon filed, see Figure 5.

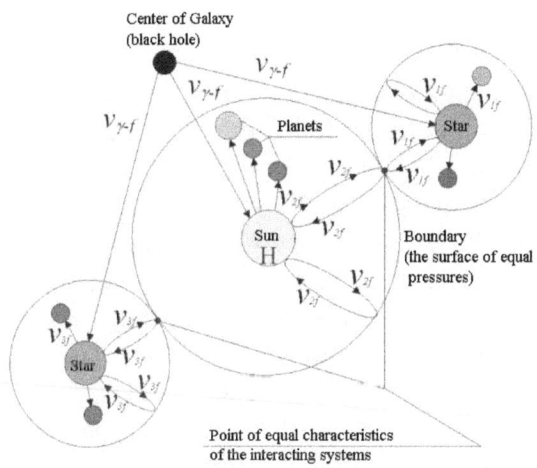

Figure 5. The diagram of interactions within the Galaxy

Let us study the energy-mass processes in the Universe in dynamics. In our model, the planets placed into the photon field of the stars have the unlimited source of the photons to produce material spatial objects from. The stellar systems, in their turn, presented in the γf field of the Galaxy have the unlimited (constantly refilled, because the Galaxy produces γf) source of energy for photon fusion. Similarly, when consuming γf_2 of the supergalaxy, the Galaxy produces its γf field where due to the fusion the stellar systems arise.

It should be borne in mind, that the energy-mass exchange in the Universe has an impact on the mutual arrangement of the spatial objects.

A star (in the case of the solar system, the Sun) is located at the center of each stellar system. A γ–photon core is produced in the Sun due to the fusion. A large quantity of the energy (the mass defect of the source photons) is released as a result of the production of (f) from γ–photons (γf) in the form of the second-level γ–photons (γf_2) $\gamma f + \gamma f = f + \gamma f_2$. This energy is superinduced from a finer level where it is packed more tightly. Moreover, the photons build in the surrounding photon field. The photon fusion in the star is followed by the generation of the objects of the material world simultaneously with the energy generation of the mass defect in the second-degree gamma photon level γf_2. Then electrons, protons, neutrons, hydrogen, and other chemical elements are generated. The similar processes take place upon the fusion inside the planets along with the matter generation and energy release in the gamma photon level (mass defect). It should be noted that after the fusion inside an object stops, a hollow is created, i. e. the object takes the shape of a hollow-body ball. The moon is a good example of this.

It should be mentioned among others that each stellar system has its own frequency (e. g. the Solar system — v_{2f}) which depends on the characteristics of the photons that such star produces; this ensures the wave propagation with this resonance frequency only within this stellar system until it reaches the boundary of its properties. This feature is connected

26

to the fact that a maximal refraction ratio of the resonance wave which is intrinsic to this system and the wave comes back to the star — the source of the latter (e. g. to the Sun), see Fig. 5.

When the wave comes back to the star, it keeps coming through the star until the maximal fraction ratio is reached. With that, the wave can not leave the body of the star and its energy remains in the star body increasing its total energy.

In the case of planets, it is also possible that a similar process takes place and this can explain, to some extent, the warming of specific layers of the planets (e. g. of the Earth).

If we study this issue in detail, using the Earth as an example, then the process probably should look like as follows. The electromagnetic wave that has approached the planet comes inside it until the fracture ratio reaches its maximum and the wave propagation stops. The energy that has dissipated in the planet body at the depth where the wave propagation stops will be spent on the "local" heating of the planet body.

A boundary (the surface of equal pressures) can develop in the areas of the potential contact of different stellar systems. Since the boundaries of the autonomous stellar systems are near spherical-shaped, the formation of the points with equal properties of the interacting systems is possible at the points of their potential contact; such points are the relatively small zones where the transit from one stellar system to another is possible without large energy consumption.

The analysis of the arrangement of the spatial objects in the Universe should study, in the first place, the interaction of the stellar systems, the planets, and etc.

While the stellar systems circle the Galaxy center in analogy with the smaller material bodies, they must be exposed to the centrifugal forces that tend to expand the Galaxy. At the same time, the consumption of the gamma photons emitted by the

27

Galaxy center contributes the balance of the forces that influence the stars. As mentioned above, the stars and their systems are exposed to the potential pressure of the gamma photon level. In addition, the stellar systems, while in the flow of the accelerating gamma photon field, interact with it and are held against the galaxy center (the black hole), i. e. a gravity interaction is initiated in the Galaxy. Each individual star, however, absorbs the gamma photons followed by the further photon fusion. Accordingly, a singular local flux of gamma photons arises and it is connected to the fusion in the black hole nd comes from the Galaxy center to each individual star; this creates the gravity directed towards the black hole as shown in Figure 6.

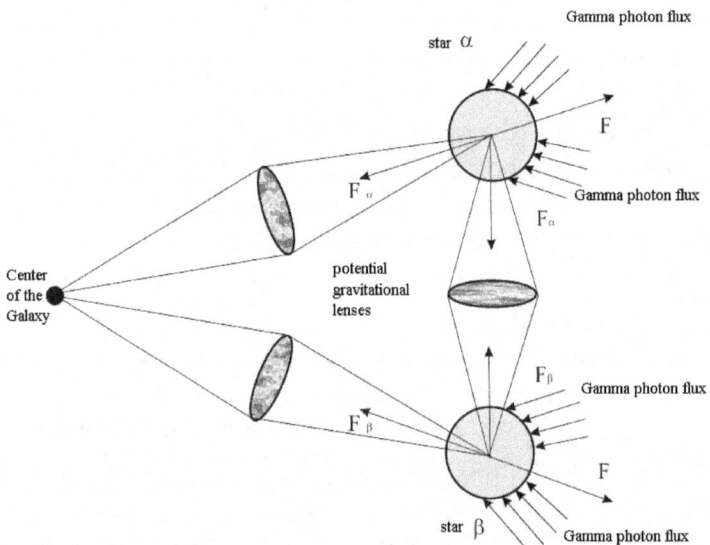

Figure 6. The forces that affect on the stars within the Galaxy

$F_{g.\alpha}$ – gravity force of the α star to the center of the Galaxy;
$F_{cf.\alpha}$ – centrifugal force of the α star;
$F_{g.\beta}$ – gravity force of the β star to the center of Galaxy;
$F_{cf.\beta}$ – centrifugal force of the β star;
F_{α} – force of holding the α star against β star;
F_{β} – force of holding the β star against α star.

At the point where a flow of any medium occurs (and the gamma photon level is medium), in accordance with the Bernoulli's equation, the local increase of the flow rate shall be accompanied also with the local decrease in potential pressure. Accordingly, the cross-section of the gamma photon level will have a relatively lower potential pressure as compared to the rest of the outer space. The total potential pressure of the gamma photon level that is effective at the surface of the star will tend to move it towards the lowest potential pressure at the surface thereof, i. e. towards $\mathbf{F_g}$ — the resulting force vector of gravity directed to the center of the Galaxy, and that, accordingly, compensates $\mathbf{F_{cf}}$ — the centrifugal force, which is opposite to it by the vector direction.

Accordingly, the mutual arrangement of the stars with their systems within the Galaxy stabilizes. It should be noted, that, as the mass of the stars rises following the fusion of the photons, the centrifugal forces that influence them also increase. In addition, the increase in the size of the stars shall be followed by the increase of fusion with the corresponding increase in $\mathbf{F_g}$ — the gravity force of the star to the center of the Galaxy. However, such increase in the stellar size shall be followed by the increasingly delayed processes in the stellar interior with the corresponding delay of the increase of $\mathbf{F_g}$ — the gravitational forces from $\mathbf{F_{cf}}$ — the centrifugal forces.

Studying the stellar interaction is also of our interest. As the stars tend to have a spherical shape, their consumption of the gamma photons shall be equal from all directions. However, this condition is disturbed in the space between two neighbor stars α and β because the gamma photons will be absorbed from this space simultaneously by the two stars. Subsequently, the flows and the reduced potential pressure of the gamma photon level are created in this space followed by the development of the above-studied attraction forces of one star to another $\overrightarrow{F_\alpha}$ and $\overrightarrow{F_\beta}$.

The planetary interaction can be explained in the same manner. The difference will only be that, in addition to the gamma photon level which the stars receive from the Galaxy's center and which surrounds them, the planets are surrounded with the photon level that comes from the specific stars. The fusion of photons to create the material particles in the planetary interiors is followed by the constant consumption of these photons from the circumplanetary space.

Due to this, the photon consumption from the interplanetary space by the two nearby planets similarly to the above described gamma photon consumption by the stars with the respective rise of the attraction forces between the planets and their mutual, energetically favorable distribution in the space of respective stellar systems.

In addition, the surfaces (lenses) must exist in all cross-sections of the fluxes of the elementary particles (the gamma photons, photons); the kinetic energies of these particles on the surfaces are at the minimum, while the potential energies are at the maximum. As the elementary particles pass this surface (also known as the Lagrangian sphere), their kinetic energy rises with the respective decrease in the potential energy.

Along with this, the characteristics of the star / planet interaction shall be studied separately; upon such interactions, the planet acts as the object of a specific mass in the gravity field of the star. In addition, the medium that accelerates towards the star center is the gamma photon level which maximally influences the center of the planet that is pressed to such a degree that it can be affected by this level in contrast to the other zones that only are penetrated thereby.

Accordingly, the planet becomes held to its star by the gamma photon level. It takes the position in the space, at which the force holding it against the star and its own centrifugal force are equivalent while their directions are opposite. It should be borne in mind, that the centrifugal force of the planet can be determined, in general, by its mass and the orbital curvature, obviously. Though, it does not seem possible to determine the intensity of compression of the center of each given planet and geometrical characteristics of this effect. We assume that, due to this, the distribution of planets in the stellar systems is not related strictly to their sizes and masses.

The material should become more understandable if we use the analysis of the well-known dependency: $E = mc^2$, which we propose to consider in an extended shape: $E \leftrightarrow mc^2$. This dependency allows giving a basis to the energy material transformation, i. e., for example, as the energy gains the known mass of some particular values, the mass of the body shall also have undergone a change. In addition, the part of the mass does not get lost; it simply transforms to the other level again by the energy. So the effect known as the mass defect can be observed.

We actually can characterize this fact as the mass defect if we operate only with the fixed parameters from the same level. But if we study the observed at the nearby levels then the same effect will be presented with a higher quality as, for example, the burning of a wooden log where its initial mass transforms not only to carbohydrate oxides but also in the heat.

The mass defect under consideration is fixed at most upon the processes, e. g. of nuclear fission and nuclear fusion. In addition, the both processes are associated with significant energy exchanges at the boundary of the levels, and, usually, they are the source of various waves. On these grounds, thus we assume that referring the fixed waves of the alignment of two "black holes" as the gravity waves can be incorrect due to the fact that the nature of these waves can be relative to the fusion.

5. Energy Transfer as the Medium Deformation

The transfer of the photon medium to the centers of fusion is anything but simple and therefore it seems worth of a separate investigation due to the fact that, in this form, it gives many characteristics to our environment.

As we have previously noted, the fusion of photons in the interior of the planet is not uniform but discrete. It follows that they also shall have a step-like arrival to the centers of fusion with short sharp accelerations and slowing-downs of the general flow of the medium with specific viscosity and elasticity. At this, it shall be noted that the viscosity of the photon medium determinately influences the character of deformation (elongation) of each separate photon towards the motion to the center of the fusion at the step-like acceleration of the flux within a short period. At the same time, the elasticity of this medium becomes determinative in a short period at step-like slowing-down of the flux and this results in the deformation of each separate photon towards the direction perpendicular to the transfer direction, in other words, it compresses towards the transfer direction. The above photon deformations are strongly connected with the properties thereof.

So, their longitudinal deformation to the transfer direction determines the creation a dipole from each photon with the opposite charges at its most distant parts. So, at this kind of photon deformation, it receives potential difference and, in general, the photon field becomes the carrier of electric intensity within this short period. In other words, within the period of interest, the electrical component of electromagnetic wave manifests itself; it has the vector directed at right angles to the transfer direction.

33

At the same time, the transversal deformation towards the photon transfer in the next short period creates the circumstances when the last becomes the carrier of the magnet component of the electromagnetic wave with the vector that circumscribes a circle around the longitudinal line of the photon transfer in the plane which is at right angles to it. The photon field becomes the carrier of magnet stress within this short period, and its vector circulates spirally at right angles to the direction of the photon flux direction towards the center of the fusion, see Figure 7.

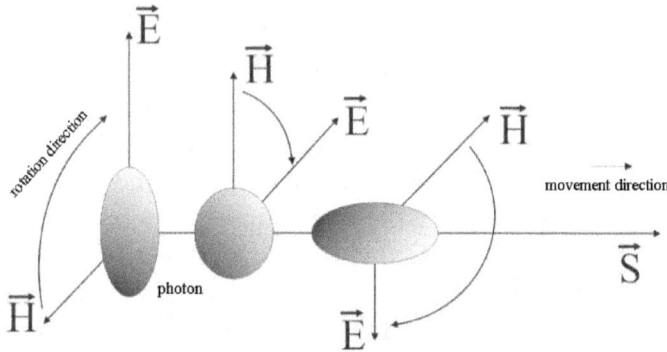

Figure 7. Photon deformation upon its motion

Accordingly, it becomes apparent that the discrete process of the fusion of the material particles from the photons in the planetary interior creates circumstances for many processes that determine the properties of the world that surrounds us. In a simplified way, if we represent the body of a planet as a living thing, then the photon field acting as an energy mass exchange would be its blood, and the pulsating process of the photon fusion can be represented as the function of its heart that ensures the circulation followed by the other complex processes. And if these processes in the living thing include, for example,

34

biochemical transformations, then everything is certainly far from simple at the space level.

The discrete matter fusion of the photons in the planetary interior, besides from acting as the energy mass exchange of the photons with the other objects in the space, ensures a range of processes that would be incorrect to call side processes.

The above-studied material proves that the photon flux moving from the source (e. g. the Sun) to the center of the fusion in the interior of a planet (e. g. the Earth) shall naturally carry electromagnetic waves. In addition, the propagation of the electromagnetic waves cannot be possible without a carrier (source) of the electromagnetic waves.

As the electromagnetic wave is a sum of the conversion of electric deformation of the magnetic deformation of the same level (or of the same field) and then the magnetic wave converts back to the electromagnetic wave, then, in a simplified way, it represents an oscillatory circuit that moves over the space from the transmitter to the receiver.

As we have already mentioned, the density of the photon flux significantly increases as it approaches the fusion area. Therefore, the electromagnetic waves that are present in the field shall also be convergent. Accordingly, the planets (including the Earth), where the fusion of the material particles (electrons, protons, and neutrons) takes place in the interior, are surrounded by the converged electromagnetic waves. The conclusion is that the flux of the photon field deforms due to the fusion which results in creating of the electromagnetic waves. The converged waves form the magnetic poles and the expanding waves are the own frequency of the planet. (The Earth, for example, has the own radiation frequency that is equal to 7.2 Hz.)

According to the current scientific views, the field state with the least energy (by absolute value) is called vacuum.

Since we do not admit the existence of vacuum due to objective reasons, we see no need for speaking on physical vacuum as a phenomenon worthy of interest.

Meanwhile, the states and properties of the photon field are of great interest for us, especially since they shall depend on the states and properties of the particles that compose the field.

First, we shall study the photon field free of any energy deposition. When it is in this state, it represents the propagation of the spherical photons that are located in the space in such a manner that the distances between their centers have the equal direction in the three planes at right angles. The spherical shape of the photons is, in our assumption, related to the fact that the sphere represents the lowest complex spatial object with the highest rigidity.

The above state of the photon field can be characterized as the "scalar state of the field" or the "scalar field". However, this state shall be regarded as ideal; in the real life, it is not practically possible due to the fact that each area of the photon field is constantly exposed to all kinds of disturbances — electric, magnetic, and so on. Accordingly, the scalar field exposed to a disturbance becomes immediately the vector field with the induction which depends on the kind of exposure, and with the vector which is opposite in direction to the vector of exposure direction.

In addition, it should be noted, that both the energy deposition and the photon field have discreteness. While it has been known since the 19th century that the energy deposition (e. g. the quantity of electricity) has the discreteness, the discreteness of the photon field calls for a separate investigation. In our assumption, the discreteness of the photon field of the specific size can be defined by the sum of all kinds of discreteness of all photons composing this field. Due to the fact that the properties of the photons set the field characteristics, the study of the photon properties is worthy of interest.

We have already noted the "scalar state of the field". Similarly to the field, the photon in the ideal state had its "scalar state", i. e. a spherical shape. The photon can remain in this state totally free of any kinds of external energy exposure (which is almost unobtainable in the real-life circumstances), or it can take this state only for a short time upon the change of its shape.

To make it simple, the change in the photon properties depending on its shape can be represented schematically in Figure 10, where the surface (1) moves in the same direction and it can affect the photons.

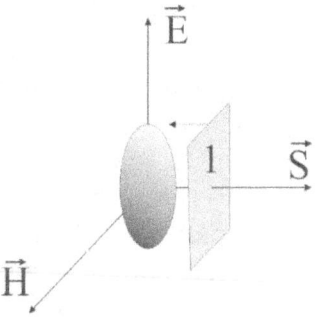

Figure 8. Scheme of the surface 1 to photons influence in the movement direction.

As we can see from Figure 8, a packing front of the photon medium before reaching the surface (1) occurs followed by the deformation of the photons that receive a slight decrease in linear sizes in the direction of the surface (1) movement and a increase in sizes by the value of $\beta = \sqrt{\dfrac{1}{E}}$ in the plane which is at right angles to the movement direction of the surface. In our assumption, this kind of deformation of the photon will be connected with the conversion of the energy of the photon exposure into the minimal discrete portion of the magnetic strain with the vector which is at right angles to the direction of exposure.

In addition, a refraction zone of the photon medium occurs before reaching the surface (1) followed by the deformation of the photons that significantly increase in linear sizes in the direction of the surface movement and a slight decrease in sizes of the cross-section of the deformed photon. In our belief, the photon, as well as the atom, has a complex structure consisting of the relatively massive, positively charged nuclear part and the relatively "light", negatively charged peripheral part which heads forward as the first area starts moving thus leaving the positively charged nuclear part behind. Accordingly, the front part of the photon, which is represented by the negatively charged part, and the rear part, which is represented by the positively charged nuclear part, when combined, create a photon dipole with a minimal discrete portion of the electric strain with the vector being at right angles to the exposure line (surface [1] movement).

The above example must not be regarded literally, as we can not create experimentally the material surface that would deform the photons due to a range of reasons; first of all, it does not seem possible to achieve the velocity of the exposure to the photons by the material surfaces. This operation, however, can be performed at the photon level in various ways.

The diagram of the photon deformation shown in Figure 8 is true only during the initial ultra-short interval because the deformed photon can have only a limited portion of energy, which, by the example of the quantity quantum, equals to $1.602 \cdot 10^{-19}$ C. After the deformation value of the photon reaches the state when the photon elasticity starts to manifest the defining properties, this deformation becomes ultimate in the longitudinal direction. In this case, the total energy of the photon is represented solely by its electrical energy part, and no magnetic properties manifest themselves. Since the quantity of photon energy is limited by its deformation value and taking into consideration the constant discrete energy conversions in the photon, the longitudinal deformation of the photon in the next short interval shall keep decreasing followed by the relevant increase in the transversal sizes of the photon.

In addition, the decrease in the electrical strain of the photon shall be followed by the increase of its magnetic strain until the total energy quantum of the photon is represented solely by its maximal magnetic energy. Accordingly, the deformation of the photon with the electric magnetic conversion of the energy will keep occurring as soon as the energy to be conducted exists. In this case, **the photons acting as the parallel inductance-capacitor oscillatory circuits fit ideally from the standpoint of the electromagnetic energy transmission over large distances with minimal losses.**

In addition to the other discreteness conditions of the processes in photon, it should be mentioned that the elasticity of the photon plays an important role as it limits the quantity of its deformation and, subsequently, the quantity of its charge, and it also plays the role of a relay that ensures the switching between the capacity mode and the induction mode of the photon and visa verse. Accordingly, the photon elasticity strongly regulates the quantity of the energy, which equals to one discrete quantum, it transmits by its capacity to limit its ultimate deformations.

The total quantity of the photons in the field capacity defines the total capacity of energy transmission by this field capacity.

On the grounds of the above, this situation actually arises when the oscillator of the photons (e. g. the Sun) produces them by the relevant "indentation" into the surrounding photon field. The indentation zone shall represent an area where the medium is packed with the relevant deformation of the photons which creates the circumstances where the maximum magnetic properties in this local quantity of the photon field can develop.

As the generator moves away from the photon oscillator, the packing degree decreases with the natural decrease in the magnetic properties. At a certain distance from the photon source, the "magnetic deformation" of the photons fades away; then they take shortly the "scalar state" due to their elasticity with the following change to the "electrical deformation". Thus an expanding magnetic-electrical wave arises, which propagates over the space from the source (e. g. the Sun that produces the photons from the gamma photons) to the consumer (e. g. the Earth that produces electrons, protons, and neutrons – material particles – from the photons). It should be also noted separately, that this wave is magnetic-electrical upon its arousal and it can only propagate due to the created circumstances, i. e. there are the oscillator that produces the photons, the receiver which absorbs these photons, and the transmission medium which can transmit the wave energy at a minimal loss. In this case, the photons that travel at a relatively low velocity and are similar to the parallel, oscillatory capacity and induction circuits, fit the most for the instant energy transport. The situation at the material level is similar to the situation when the high-frequency pressure waves arise in the common flow of the pumped liquid due to certain reasons; these waves can transport the most of the energy over the large distances at an extremely high velocity.

Both for the wave photon flux and for the pressure wave, the corpuscular and wave streams are concurrently present in the flow of the actually pumped liquid. However, for the hydraulic impact, the propagation of the wave of high-frequency disturbance in the general flow of the actually current liquid is regarded as self-evident, while the magnetic-electric (electromagnetic) wave in the photon flux is followed by the continuing scientific disputes regarding the corpuscular and wave nature of light. In our belief, the light represents an electromagnetic wave in the moving photon flux at a relatively low velocity. Accordingly, the assumption of the existence of the ether medium corresponds to all issues of the dualism regarding the "wave — particle" nature of light that has aroused among the scientists.

After consideration of the features of the magnetic-electric wave within the photon field at the photon oscillator and its propagation over the space, how it approaches the receiver should be of interest. As we have mentioned above, the planets can act as such receiver in the stellar systems (e. g. the Earth planet in the Solar system); when the wave approaches it, a refraction area develops in the flow due to the photon consumption in the planet interior to produce the material particles. That is why the photons that approach the planet will have an "electric deformation", which, initially, changes to the "scalar state" as it moves away, and, finally, to the "magnet deformation".

Since the high-frequency energetic waves from the star propagate in the photon fluxes that are converging to the planet center, they will be seen, for example to a witness on the Earth surface, as converging electromagnetic waves in contrast to expanding magnetic-electrical waves that would be seen as expanding waves from the Sun surface. This is schematically shown in Figure 9.

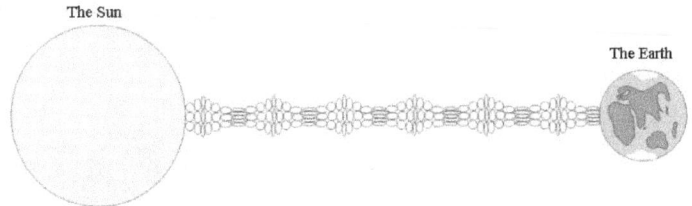

Figure 9. Scheme of the propagation of the energy wave
within the space

It should be noted that the wave propagation is possible only in an elastic medium where alternate compression and refraction of the medium followed by the simultaneous acceleration and slowing-down of the particles that build it. **No acceleration — no force that affects the medium.**

The gravity of spatial objects is determined by the nature of the fusion that takes place in these objects. The electromagnetic radiation from these objects that evolves upon the fusion takes place at various levels, see Figure 10.

The gravity process is the actual movement of the photon field to the fusion center at acceleration. It is followed by the electromagnetic radiation of a strongly defined frequency.

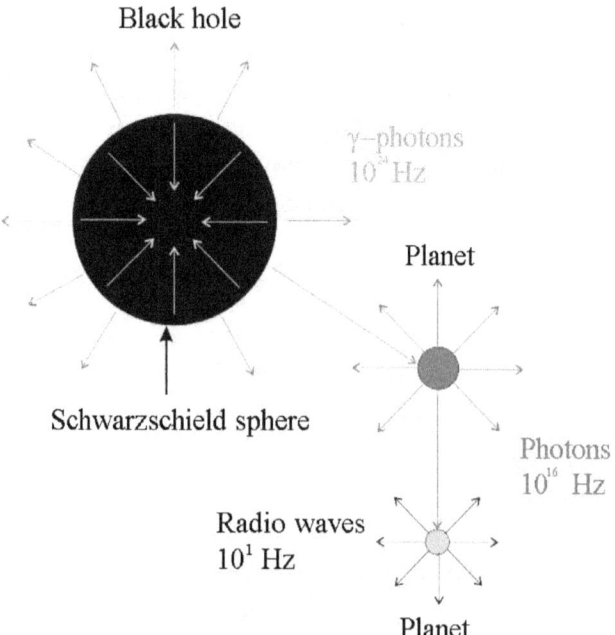

Figure 10. The gravity and radiation frequencies of various space objects

The gravity develops as an interaction of the objects within the ether medium that itself takes part in energy and matter conversions. The gravity process is the interaction of the object with the photon field.

We have studied the issues of gravity, electromagnetic wave, internuclear interactions that were demonstrated by the example of the formation of chemical elements [https://www.youtube.com/watch?v=NZAJXfgSr1I], as well as of the interaction of spatial objects. The issue of charge still remains uncovered. In our belief, the elementary charge is a lowest-stable aggregation of the ether elements with a higher density (minus) or a relatively lower density (plus) that is involved in a vorticose movement — right or left respectively; it simultaneously forms a closed energy system.

6. Conclusions:

1. **Ether is a multilevel substance**
2. **All particles have a complex structure**
3. **All interactions are created by the deformation of ether**
4. **Interaction laws can be described by the same equations. The field strength at all ether levels decreases proportionally to the square of the distance between the objects, Coulomb forces or gravity forces in accordance with Newton.**
5. **The Universe itself has originated from the fusion, which, similarly to black holes, stars, and planets, has initially formed at an infinitesimal point. Thus, we have the Universe expansion. There were no packed Universe in an infinitesimal quantity but rather there was an initiation of the synthesis of matter from the field.**

And there the question of what is time is still remaining. **Time is a change in energy state of the object.** It depends on the acceleration at which an object or a field comes through it causing an increase in the energy density of the object. The higher the density is, the slower time passes. Accordingly, **time is the function of the energy density of the object (subject).** A good example of this is the neutron — it can stay for billion years in atomic nucleus while it stands only for 600 seconds in the free state. Any clock, whether electronic or mechanical, slows down in mines but it will be running ahead of time in the orbital space station, which is an established fact.

7. The Examples of the Existence of Ether

The advances of the contemporary science with regard to the registration of gravity waves give a proof of the fact that such waves, equally to any other waves, can propagate the energy over astronomically large distances in an effective way, and, moreover, to exist during extremely long intervals. However, the report materials regarding this issue say nothing about the medium which made the propagation of the waves possible. Accordingly, in the light of the circumstances of propagation of the registered waves when their energy has not been dissipated for a long time (has not changed to the internal energy), and even from the common point of view of contemporary scientific approach, all the above said shall occur in an absolutely elastic medium. That is why it should be noted that if a continuous medium has elastic properties then the motion of the points at one location of the medium (in the source) well lead to the propagation of this motion at a certain velocity in the shape of an elastic wave. In addition, if the medium has only its cubical elasticity, then only the longitudinal waves (liquid, gas) can propagate; and if it has also a shape elasticity, then the transverse (shear) waves are also possible. This just indicates that the medium of a multilevel ether has all the above-specified properties because it can propagate not only longitudinal waves but also transverse (shear) waves. Cumulatively, the first and the last create electromagnetic waves. In addition, both the longitudinal and the transverse waves can exist independently of one another.

It should be noted separately, that the propagation of any waves without a medium (in the vacuum) is not possible. In accordance to the above said, the study of the medium where the propagation of the waves of all kinds takes place shall be considered as a separate matter.

First of all, it should be noted, that any registration instrument has a strongly limited measuring range. While the currently available measuring tools allow registering the neutrons, protons, and electrons individually by the specific particle characters, the photons, for example, the photons in a thinner level, cannot be registered by the current tools as the individual particles; that does not mean, though, that they do not exist. For example, if we had made a hot-water injection to the water, then the volume and mass of the ocean would change, the microwaves would arise, the temperature in ocean would change, but no tool can ever register all this. Accordingly, all the energy exchange processes that continuously take place and the quantity of which is countless use a thinner level as an endless source of energy for one part, and for the second part as utilizer of any excess energy. The above said gives us the ground to conclude that **energy is a sum of the particles in the thin-layer ether that are involved in energy exchange processes and remain in an unstable state.** This leads us to a conclusion that **an individual Planck's constant shall be available for each ether level.**

To date, the capabilities are available that ensure the registration of the propagating waves of electromagnetic radiation and the determination of light pressure to which the reflection and absorption bodies, particles, as well as independent molecules and atoms, are exposed. In addition, while the wave properties of light are regulated by the wave laws that involve elastic media, it currently is not possible to register the corpuscular properties of light for each individual component of the particle due to a lack of capability to register these individual particles at the moment. Meanwhile, the sum pressure of the particle flux had been registered back in 1899 by P.N. Lebedev.

The above material gives a proof that, firstly, the propagation of the registered gravity waves shall involve some medium. However, the issue of whether it is correct to regard these waves as the gravity waves can naturally arise.

In fact, there can be no doubt in the registered near collision of two black holes followed by their merging. The above merging shall very likely occur at least in two levels followed by the changing from one to another. As we have mentioned above the transition of the matter (energy) from one level to another shall be followed by the mass defect. While previously there was no need to study this effect separately, now such need has aroused.

In our beliefs, it is the presence of the ether in the stellar space that explains the rotational direction of the planets, e. g. of the Solar system. It is known that the planets and the most of other objects circulate around the Sun in the same direction with it (counter-clock, from the Sun north pole view). The most of the planets rotate around their axes in the same direction as they circulate around the Sun. The exceptions are Venus and Uranus; what is more, Uranus circulates literally laying on its side (the axial tilt is 90°).

It seems quite probable that there should be something that has affected the rotation of Venus and Uranus from outside, or the cause is to be sought in their structure patterns, their arrangement against the other space objects, and etc. This issue should be the subject of a separate study. For us, the explanation of the rotation of the most planets is chiefly of importance at this stage. It would be a good place to start a detailed study with the assumption that, firstly, it is reasonable to study the contingently even motion of a spherical body (a planet similarity) in the vacuum, see Figure 11).

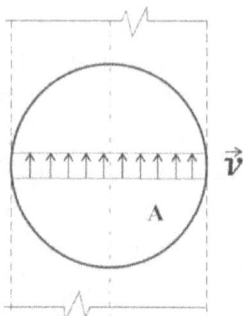

Figure 11. The diagram of contingently even motion of a spherical body A in the vacuum, where $\vec{v}$ — is the travel velocity of the separate points of the body A

Figure 11 proves that all points of the body A travel over the space free of any exposure. There is no single reason why the body A would change its position as the diagrams of the velocity are equal for all points of the body.

Next, let us study the contingently even motion of the spherical body in a circle with R radius, see Figure 12.

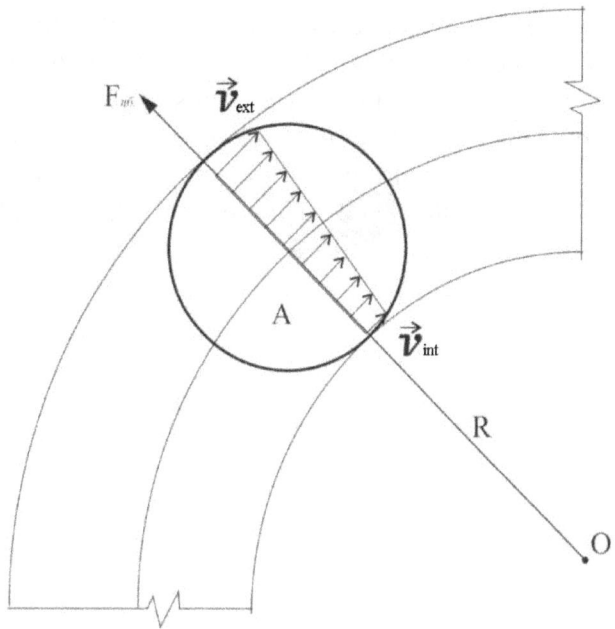

Figure 12. The diagram of contingently even motion of the spherical body A in a circle with R radius,

where: $- \overrightarrow{V_{int}}$ is the value of the travel velocity of the points of the body A near the center of rotation, $\overrightarrow{V_{ext}}$ — the travel velocity of the points of the body A that are more remote from the rotation center $\overrightarrow{F_{cf}}$ — the centrifugal force.

The different velocity values of the various points of the spherical body A, which moves in the circle with R radius, are not only interrelated with the arising centrifugal force $\overrightarrow{F_{cf}}$, but

51

they also create an unbalance which is illustrated in the diagram of velocity of the body A, see Figure 12. Due to the fact that everything in the surrounding world trends to reach a balanced, energetically beneficial state, then, from the merely theoretical mechanics view, the occurrence of the counter-clock rotation of the body A at the rate 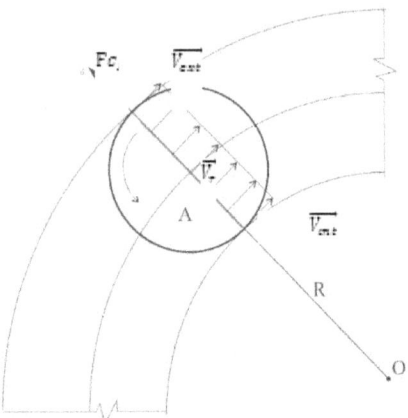 in Figure 13 would lead to the alignment of the velocity values of its various points on the diagram.

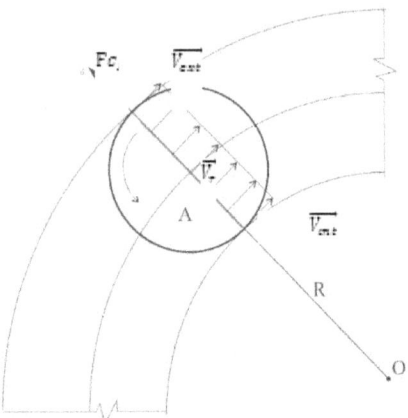

Figure 13. The diagram of a variant of the alignment of velocity values of the various points of the spherical body A upon its movement in the circle with R radius

In addition, it is easy to determine the rotational velocity of this spherical body because, as is well known, the change in the velocities of individual points on the diagram is of linear character. To determine the velocity of the rotation of the body under consideration, it would be enough to sum up the values of the velocities of and and divide this sum by 2. Thus, we obtain the following expression:

$$V = \frac{1}{2}\left(V_{int} + V_{ext}\right)$$

In addition, we shall note tha $\overline{V_{ext}}$ upposed direction of the rotation of the spherical bodies in th.. vacuum is diametrically opposed to the actual direction of rotation of the most planets in the Solar system; this, in its turn, leads to a conclusion that the motion of the planets occurs not in the vacuum, but in the ether medium.

That is why we will try to study everything that the planetary bodies (near spherical-shaped), e. g. the Solar system can encounter upon their circulation around the star (the Sun) in the ether media. This is schematically shown in Figure 14.

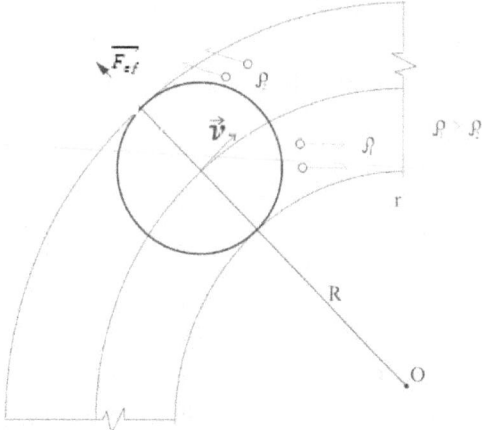

Figure 14. The diagram of the ether medium / planet body interaction along the planet movement around the Sun

As we can see from Figure 14, the exposure of the points in the planet body that are more remote from the rotation center to the ether medium occurs mostly tangentially in contrast to the almost frontal exposure of the body points that are located nearer to the rotation center. In addition, the ether density (ρ) is higher towards the star (the Sun) and, therefore, the planet points that are nearer to the rotation center shall constantly undergo a higher resistance due to this. Accordingly, the most exposure of the points in the planet body that are located nearer to the rotation center shall naturally cause the revolving of the planet around its axis in the direction towards the rotation around the center (the Sun), which we can objectively observe. This fact also proves that the planetary rotation, e. g. of the Solar system, occurs not in the vacuum but in the ether medium.

A separate consideration of the prerequisites of the generation of centrifugal force is also of our interest.

From the standpoint of the current scientific approach, the centrifugal force shall be regarded as a part of fictitious forces of inertia which is added upon conversion from the inertial reference system to the non-inertial, properly rotating system. In addition, it is generally thought that, in this way, Newton's laws can be still applied within the obtained non-inertial reference system upon the estimation of the acceleration of bodies through the forces balance. Often, especially in the technical literature, they change implicitly to the non-inertial referential system, which rotates with the body, and refer to the inertia law manifestations as to centrifugal force acting on the side of the body, which moves by a circular trajectory towards the connections that caused this rotation, and estimate it by definition as equal to the centrifugal force in modulus and always directed to the opposite side.

In our point of view, it is not only such definition fails to give a clear, brief explanation of the actually registered physical effect, but, moreover, the use of such non-relevant, for the exact science, terms, such as "fictitious", "implicitly", and so on, obscures even further the explanation process itself.

Let us try to explain the originating of the centrifugal force operating with common concepts upon analyzing, see Figure 15.

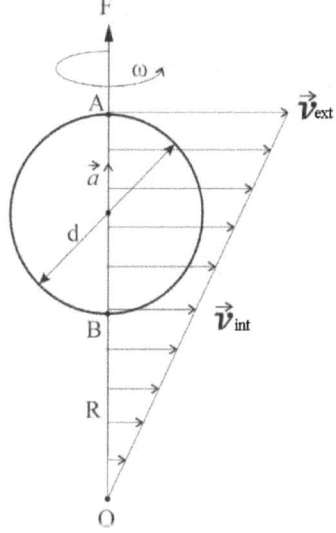

Figure 15. The diagram of originating of the centrifugal force by example of the rotation of a ball-shaped hard body of d – diameter and m – mass along the circular orbit of R – radius at ω — an angular speed

Where:

 – the linear velocity of the body points on the outer side of the rotation;

 – the linear velocity of the body points on the internal side of the rotation;

55

α   – acceleration;

     – centrifugal force.

The epure of linear velocities of various body points in Figure 15 shows that the linear velocity in the point A is higher than in the point B, therefore, if we operate with the concepts applicable to the straight-line motion, then it would be logic to say that an acceleration must act in the direction with a higher linear velocity.

The product of the mass of the body by its acceleration can be presented as the force directed towards the acceleration vector. This force directed radially away from the rotation center to the periphery is the centrifugal force.

8. PS.

Having considered all materials presented herein, we are led to the conclusion that the Universe is a complex self-developing system while the man is only a part of the system, but the part will never be more complex than the whole. Accordingly, the human mind is only a part of the Universe mind!

Reference List

1.   Kontseptsii sovremennogo yestestvoznaniya. Konspekt lektsiy [Concepts of Modern Natural Science. A Summary of Lectures] (In Russian) / Karpova, T. V. - M.: AST, 2010. 160 p.

2.   Einstein, A. Sobraniye nauchnyky trudov v chetyrekh tomakh [A Collection of Scholarly Works in Four Volumes] (In Russian), M.: Nauka, 1965–1967

3.   Matveyev, A. N. Atomnaya fizika. Uchebnoye posobiye dlya studentov vuzov. [Atomic Physics. Tutorials for Students of Higher Education] (In Russian.). M.: Vysshaya shkola. 1989. 439 p.

4.   Sheshukov V. V., Groh, V. Ya., Novyy vzglyad na stroyeniye materii [A New Approach to the Structure of Matter] (In Russian), Dom Pechati. Chelyabinsk. 2011. 104 p.

5.   Ivanov, I. S. Elementy bolshoy nauki / Neitrony raspadayutsya i s izlucheniyem fotonov [The Elements of Big Science / Neutrons Break Down also with the Photon Emission] (In Russian) http://elementy.ru/news/165038

6.   Yavorskiy, B. M. Spravochnik po fizike [The Physics Handbook] (In Russian) / Yavorskiy, B. M. and Detlaf, A. A. M.: Nauka, 1991. 942 p.

www.ingramcontent.com/pod-product-compliance
Lightning Source LLC
Chambersburg PA
CBHW061219180526
45170CB00003B/1072